CONTINENTS

Library Edition Published 1990

Published by Marshall Cavendish Corporation
147 West Merrick Road
Freeport, Long Island
N.Y. 11520

Printed in Italy by Imago Publishing Ltd

Designed and produced by AS Publishing

Library of Congress Cataloging-in-Publication Data

Mariner, Tom
Continents / by Tom Mariner,
p. cm. – (Earth in action)
"A Cherrytree book."
Includes index
Summary: Examines the physical character of the earth's seven continents.
ISBN 1-85435-195-8
1. Continents – Juvenile literature. [1. Continents.]
I. Atkinson, Mike, ill. II. Title. III. Series: Earth in action
(New York, N.Y.)
G133,M346 1989 89-17285
551.4'1–dc20 CIP
AC

·EARTH·IN·ACTION·

CONTINENTS

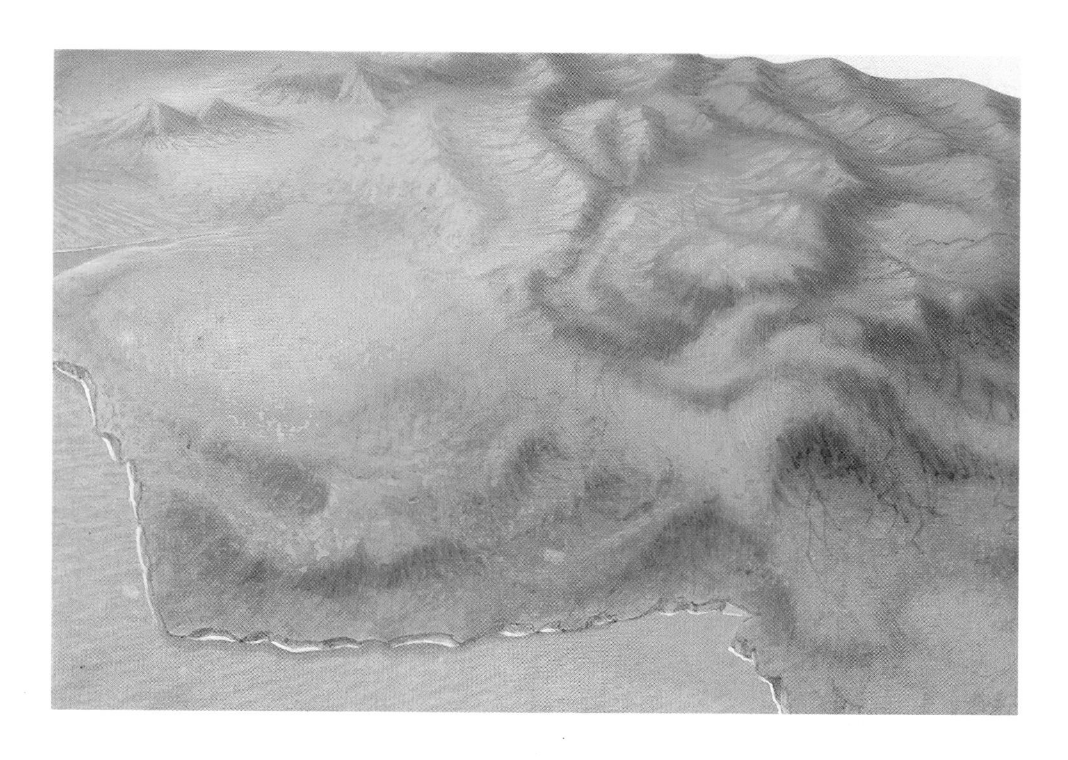

Tom Mariner
Illustrated by Mike Atkinson

MARSHALL CAVENDISH
NEW YORK · LONDON · TORONTO · SYDNEY

The Seven Continents

Water covers about seven tenths of the earth's surface. The rest is dry land. There are seven main land masses, or continents. In order of size, they are Asia, Africa, North America, South America, Antarctica, Europe and Australia. The continents are huge, unbroken land areas surrounded, or largely surrounded, by water. They also include nearby islands. For example, North America includes Greenland, the world's largest island.

Asia and Europe have a long land border, which runs along the Ural Mountains and the Ural River, and then through the Caspian Sea and the Caucasus Mountains. Because of this long land border, Asia and Europe are sometimes considered to be one continent, called Eurasia.

Africa is joined to Asia along the short Suez Canal, while North and South America are linked by a thin strip of land called the Isthmus of Panama. Antarctica and Australia, the only country that is also a continent, are completely surrounded by water. Australia, New Zealand and Papua New Guinea are together often called Australasia.

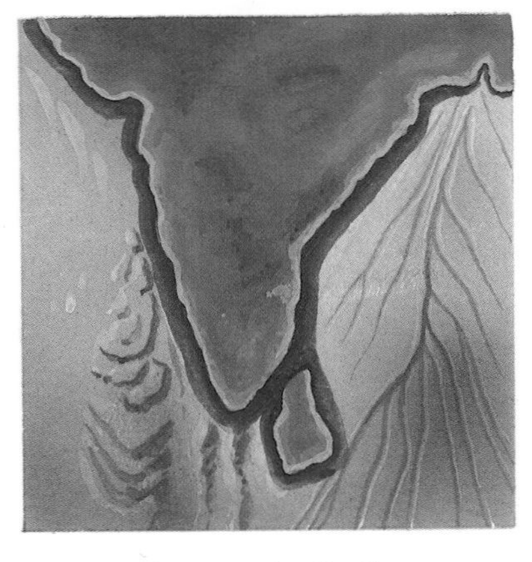

In this picture, the Indian Ocean has been drained away to show the continental shelf off the coast of India. A high part of the shelf forms the island of Sri Lanka.

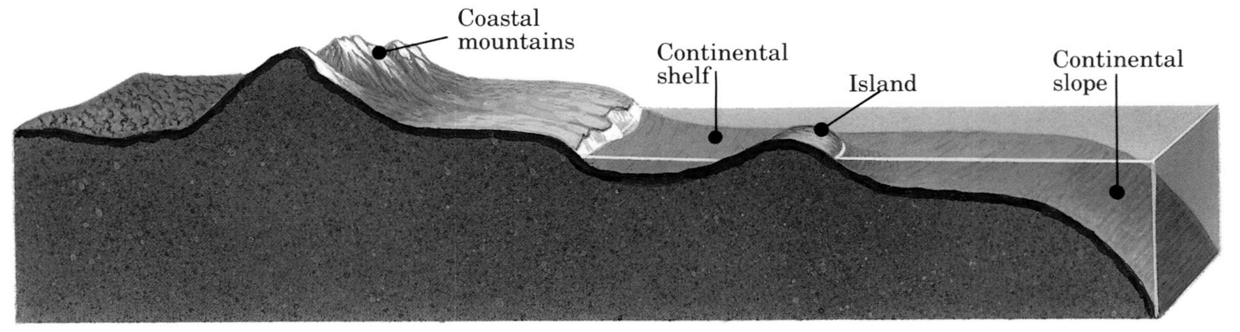

The edge of a continent: the continental shelf is part of the continent, even though it is under the sea. The shelf ends where the continental slope drops to the ocean floor. Offshore islands are simply high parts of the shelf.

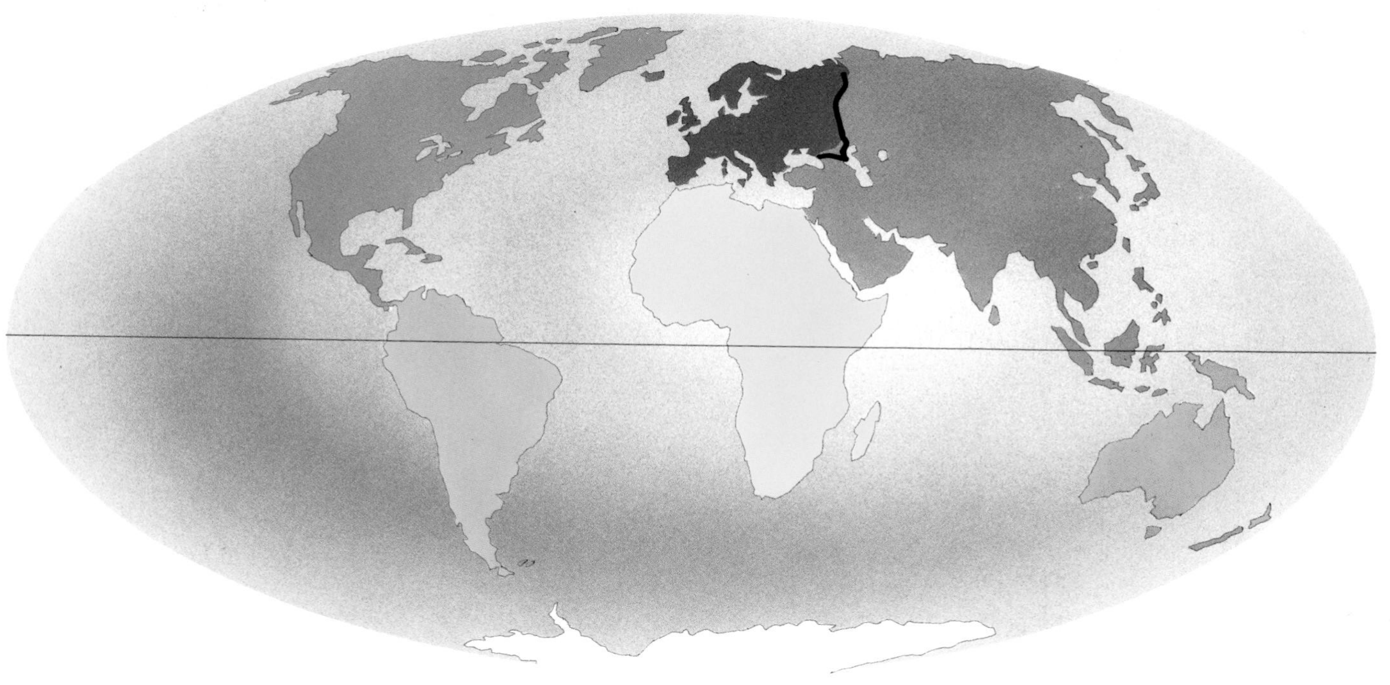

Shelves and Slopes

Around the shores of continents are shallow seas. These seas cover gently sloping areas called *continental shelves*. The continental shelves slope gently to a depth of about 650 feet (200 m). They end at the top of the *continental slope*. This steeper slope leads down to a deep ocean floor, which is called the *abyss*. The true edge of a continent is not, therefore, its coastline, but the edge of the continental shelf.

The width of continental shelves varies greatly. Off the west coast of South America the shelf is narrow, but off the coast of northwestern Europe, it extends several hundred miles out to sea. The higher parts of continental shelves stand up above sea level to form islands, such as Great Britain and Ireland off the coast of mainland Europe.

The seven continents: land covers about three tenths of the earth's surface. The northern hemisphere contains more than twice as much land as the southern.

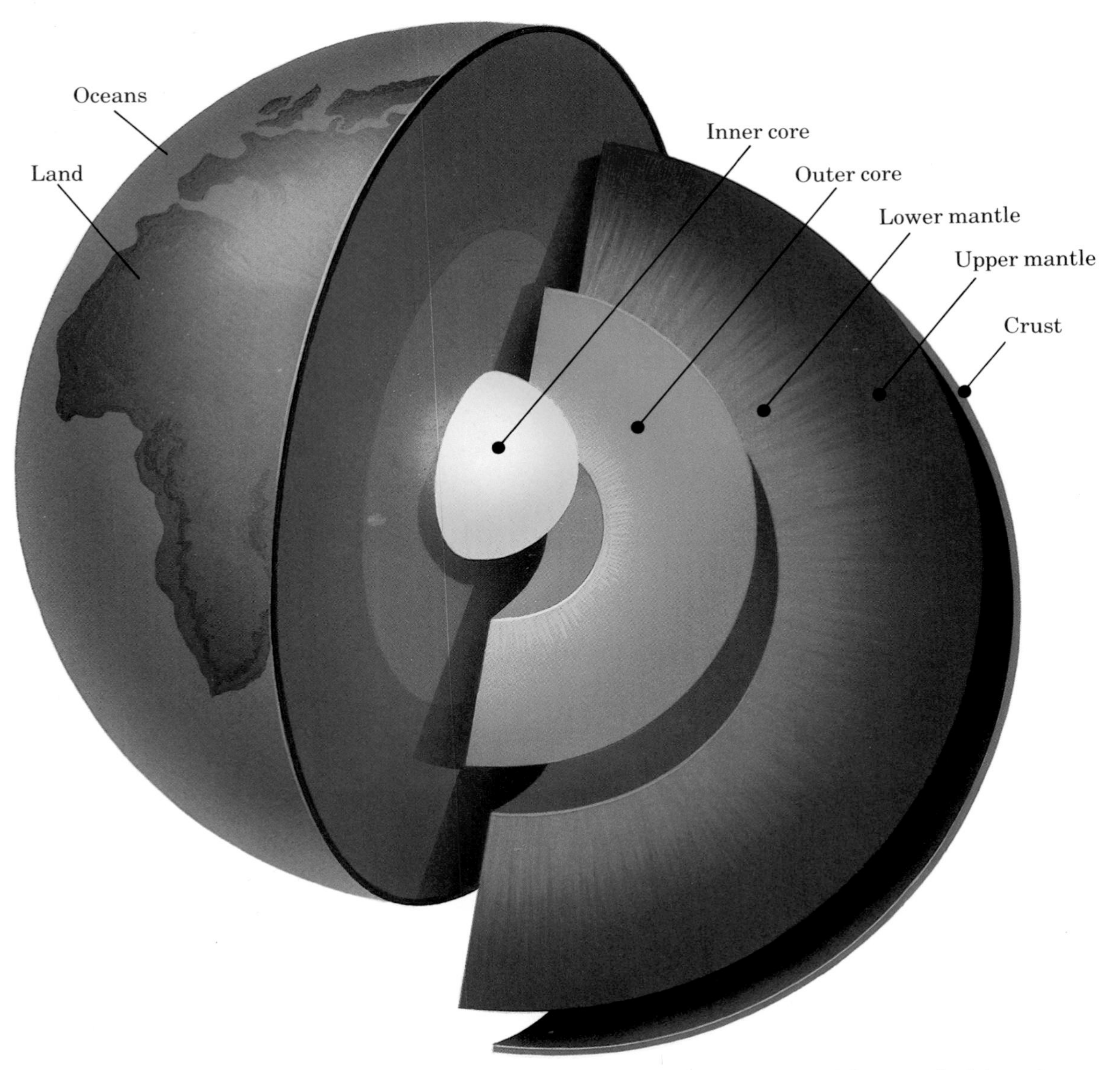

Core At the center of the earth is a solid ball of nickel and iron. Surrounding the inner core is the outer core, a layer of dense, molten rock.

Mantle This layer forms over three quarters of the earth's bulk. The rock becomes steadily less heavy toward the crust. Part of the upper mantle is molten.

Crust The crust is thin and light: the rock under the oceans (ocean crust) is heavier than the rock which forms the continents (continental crust).

The Structure of the Earth

The earth was formed about 4,600 million years ago. It was a ball of dust and gases, so immensely hot that they formed a molten mass. As time passed, heavier materials sank towards the center of the planet, while lighter ones moved outward, forming several layers.

As time passed, the earth cooled, and the surface hardened into a *crust* of solid rock. Steam rose into the atmosphere and formed thick clouds. The rain from these clouds filled hollows in the surface of the planet and formed the ancient oceans. The higher land became the continents.

If you put a pan of water on the stove (above), the water nearest the heat becomes hot and rises. Cooler water from the top sinks to take its place. Then, it too rises, and so on. As the water gets hotter, it continues to circulate in currents in the saucepan. Molten rock in the upper mantle behaves the same way. The currents in the magma cause the land above to move.

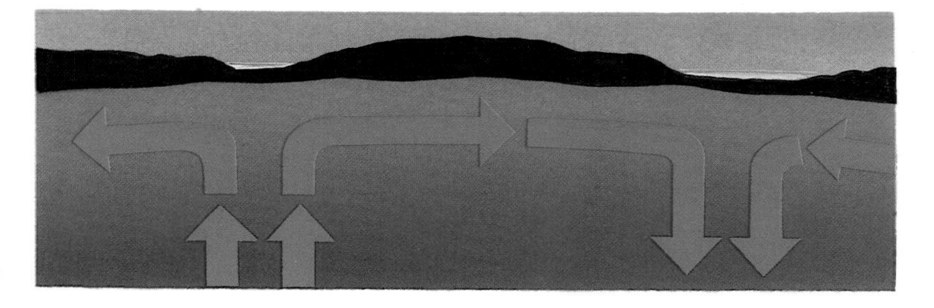

The distance from the surface of the earth to the center is about 4,000 miles (6400 km). The earth's crust is thin – only 3½ miles (6 km) deep in places under the oceans. Below it is the *mantle* which is about 1,800 miles (2900 km) thick. In the center is the earth's *core* which is about 4,325 miles (6920 km) in diameter.

The top of the mantle is solid, like the crust. But between 45 and 65 miles (70 and 100 km) from the surface, the rocks are close to melting point. These rocks flow slowly in rotating currents, like a boiling liquid, but much more slowly.

Most rocks are hidden from view by soil and vegetation, but you can see them in cliff faces, in canyons or in quarries. Sedimentary rocks are often arranged in layers, or *strata*, like those in this cliff.

The Earth's Crust

For many millions of years after the earth formed, its surface was molten. As the earth cooled, parts of the surface hardened into a crust, but these hardened parts were later broken up and remelted. This explains why the oldest rocks found on earth are about 3800 million years old. Most of the rocks formed before that were destroyed.

Igneous Rock

Rocks that are formed when molten material, called *magma*, cools and hardens are called *igneous* rocks, from the Latin word *igneus*, meaning fire. Igneous rocks form today when a volcano erupts and molten rock pours out over the surface. When magma reaches the surface, it is called lava.

Sedimentary and Metamorphic Rocks

After the crust had formed, fierce storms raged over the earth. Rain and running water wore away the first land areas, which were made of igneous rocks. The wearing away of the land is called *erosion*. The worn material, or *sediment*, was carried by rivers into the seas where it piled up in layers. It slowly became compressed into hard layers of *sedimentary* rock.

The earth is always changing. Molten rock in the upper mantle moves parts of the crust around, squeezing rocks together into mountain ranges. Hot magma forced upward into the crust also bakes the rocks. Pressure and heat change igneous and sedimentary rocks into new *metamorphic* rocks.

KINDS OF ROCK
Igneous rock forms from molten rock. Some comes from lava that pours out of volcanoes and hardens. Some hardens underground.

Sedimentary rock forms from sediments. These are particles of other rocks that have been eroded. They are carried in the form of sand and mud to the sea by rivers. Under the sea, they pile up in layers and slowly harden into solid rock.

Metamorphic rocks are rocks that have been changed by heat or pressure into another kind of rock. The heat comes from the hot magma under the ground or from the pressure of earth movements.

The rocks of the continental crust are rich in two elements, silicon and the lightweight metal, aluminum. Continental crust is called *sial* from the first two letters of these two words. Oceanic crust is rich in silicon and magnesium, a heavier metal than aluminum. The word for ocean crust is *sima.*

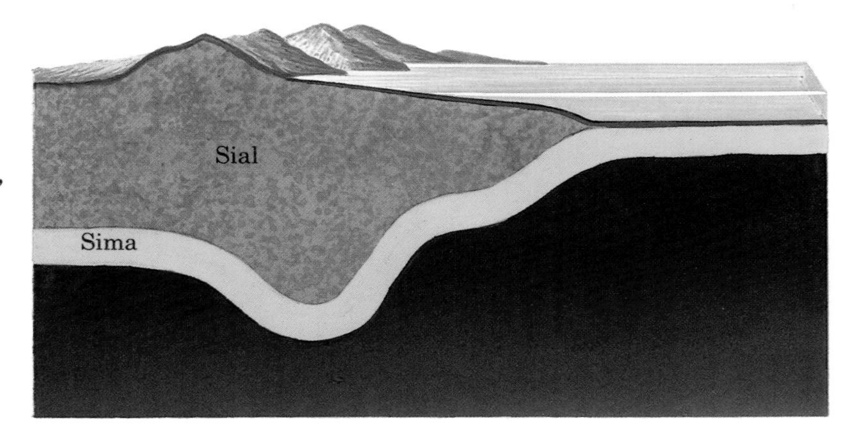

Drifting Continents

As soon as the first maps of the world were drawn, it must have been obvious that some continents would fit together like pieces in a jigsaw, if they were not so far apart. It was not until 1915 that anyone seriously proposed that the continents had once fitted together.

A German scientist, Alfred Wegener, made the suggestion and pointed to the evidence. Besides the apparently perfect fit of South America and Africa, Wegener noticed that if you fitted the continents together, mountain ranges of the same age and type would be next to each other. Wegener also discovered that areas of South America, Africa and Australia had once been covered by ice. Had they then once been at the South Pole? Wegener thought they must have been, but other people scoffed.

Proof

Scientists know that different kinds of animals evolve in different parts of the world. Animals in Australia are not like those in Europe or South America. And yet fossils of certain animals were found in different continents. This convinced Wegener that the continents had once all been joined together, but since he was neither a geologist nor a paleontologist (student of fossils), nobody took him seriously. How could continents move? What force could thrust solid land masses so far apart?

It was not until after Wegener was dead that scientists acknowledged that the continents had moved, and found conclusive evidence to prove it.

EVIDENCE OF DRIFT

About 200 million years ago, all the continents were joined in one huge land mass called Pangaea, which then split up again. The southern part of the continent was also once at the South Pole.

Magnetized particles in old rocks show that the continent gradually moved over the South Pole between 300 and 200 million years ago.

Geologists have also found evidence of glaciation – ice covering the land – in Africa, as far north as the Sahara, which is now desert.

Rocks on the west coast of Africa and the east coast of South America are exactly alike.

Identical fossils of animals and plants, which could not have evolved on separate land masses or reached them by sea, have been found. They include those of a land reptile called Lystrosaurus, *a freshwater reptile called* Mesosaurus, *and plants called* Glossopteris.

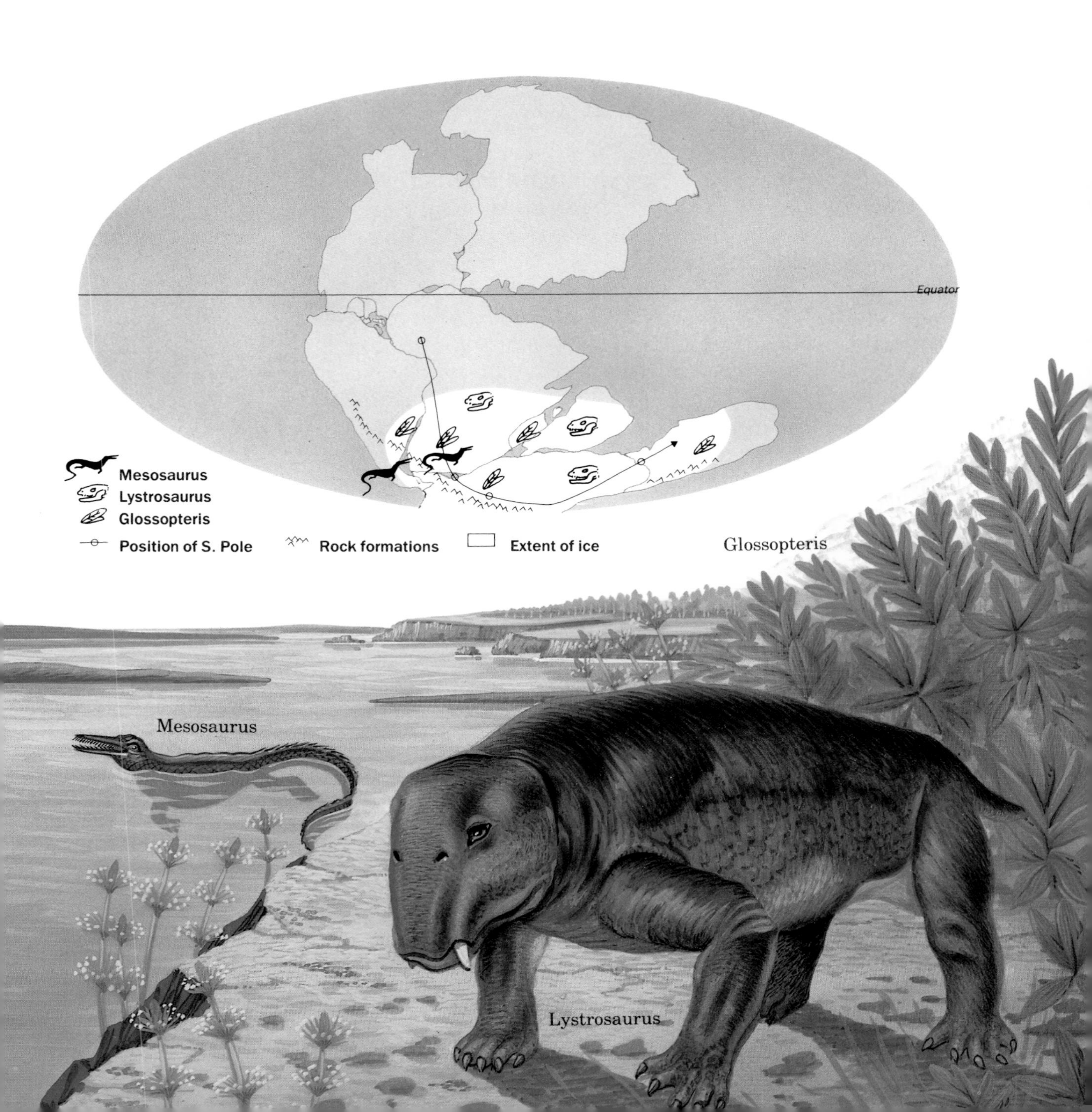
Equator
Mesosaurus
Lystrosaurus
Glossopteris
Position of S. Pole
Rock formations
Extent of ice
Glossopteris
Mesosaurus
Lystrosaurus

Secrets of the Ocean

The evidence for Wegener's theory lay under the oceans. When the ocean floor was mapped in the 1950s and 1960s, scientists discovered long undersea mountain ranges, called ocean ridges. Deep rift valleys ran the length of the middle of the ridges. Volcanoes, were scattered alongside the mountains. In places, the volcanoes surfaced as islands. Elsewhere, there were long deep trenches, 6 miles (10 km) below sea level.

The scientists worked out the age of the undersea

The ocean ridges are the world's longest mountain ranges. The rift valleys in the centers of the ridges are composed of young rocks that are hotter than the rocks on either side. This means that new crustal rock is being formed along these valleys. Alongside the deep ocean trenches where earthquakes are common, there are volcanic island chains.

rocks and found that the rocks in the rift valleys were the newest, or youngest, in the oceans. Away from the ridges on both sides, the rocks got older and older – although not as old as those on land. No rocks on the ocean floor are more than 200 million years old.

The ridges were also unstable. Not only did volcanoes erupt continually, there were also frequent earthquakes. There were earthquakes along the deep ocean trenches, too. Clearly, the rocks must be moving to provoke such activity.

SMOKERS
In places along the ocean ridge, super-heated water gushes from deep clefts in the ocean floor. The water is colored by minerals dissolved from the newly-formed rock. These boiling springs are called smokers.

Moving Plates

From studies of the ocean floor, scientists came up with a new theory called *plate tectonics*. The earth's crust is not a continuous unbroken shell, as people once believed. It is split into large, rigid blocks called plates. They are moved around by currents in the molten rock in the upper mantle. The rift valleys in the ocean ridges are the edges of plates. Here, new crustal rocks are continuously being formed as two plates move apart.

Plates move apart along ocean ridges, making the oceans wider. They are called spreading ridges. When plates collide, one plate is pushed under another plate along a trench, or *subduction zone*. The descending plate is melted and destroyed.

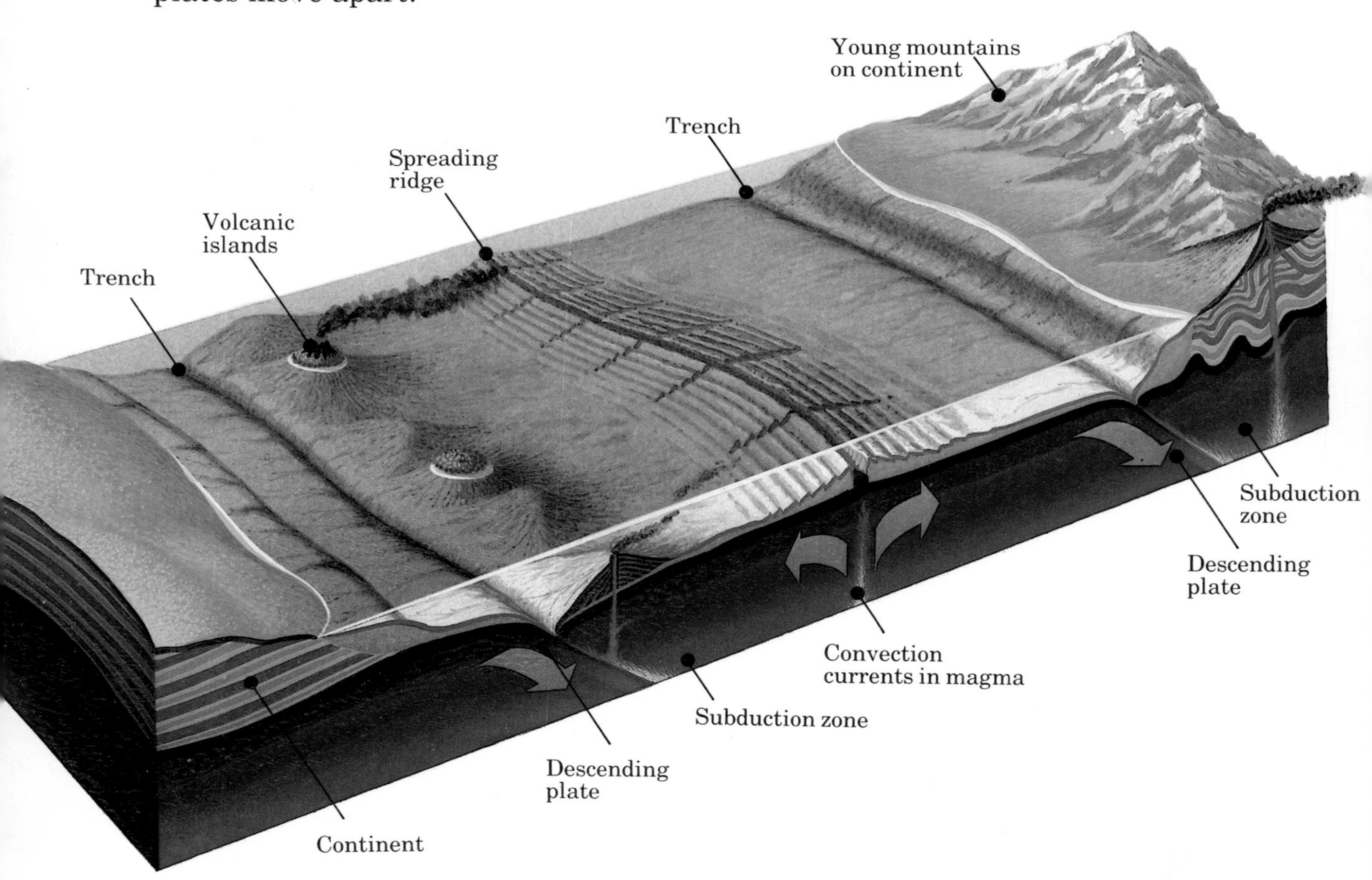

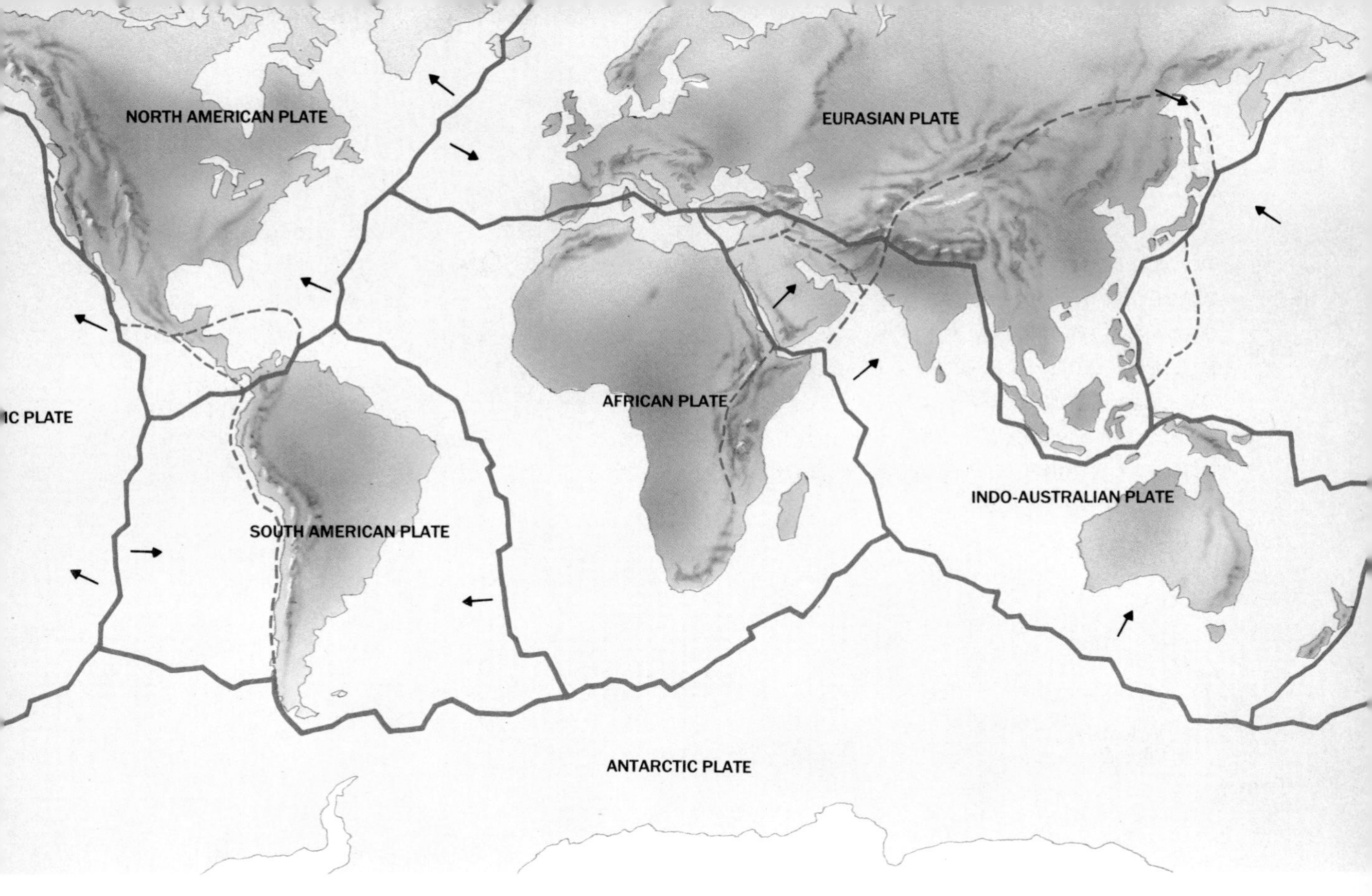

Melting Rocks

The deep trenches are also plate edges, but crustal rocks there are being destroyed. One plate is pushed beneath another plate into the interior of the earth. The edge of the descending plate grows hot and melts, creating magma. Some of this magma rises upward to reach the surface. Chains of volcanic islands in the sea have been formed this way. Volcanoes on land are also the result of magma welling up through the continental rocks and seeping, or erupting, through cracks in the surface.

The earth's hard, outer shell is split into seven huge plates and several smaller ones. Plate movements have been going on throughout most of the earth's history and are continuing today. The arrows indicate the direction of their movement.

Earthquakes

Most earthquakes occur when rocks in the earth's hard outer shell move along *faults*, or cracks. Plates do not move smoothly. Their edges are jagged and become locked together. Gradually, strain builds up as the currents of molten rock inside the earth continue to push and pull at the overlying plates. Eventually, the strain becomes so great that the plates lurch forward in a sudden movement. This movement makes the ground shake.

Earthquakes can occur anywhere. But the most severe earthquakes occur near the plate edges. Earthquake zones follow the ocean ridges and the ocean trenches. They also follow a third type of plate edge, called a *transform* fault. Along transform faults, such

The diagrams (opposite) show what happens when rocks move along faults. Some rocks are pushed up to form highlands bounded by steep edges, called *fault scarps*. Sometimes, a block of land sinks down to form a rift valley.

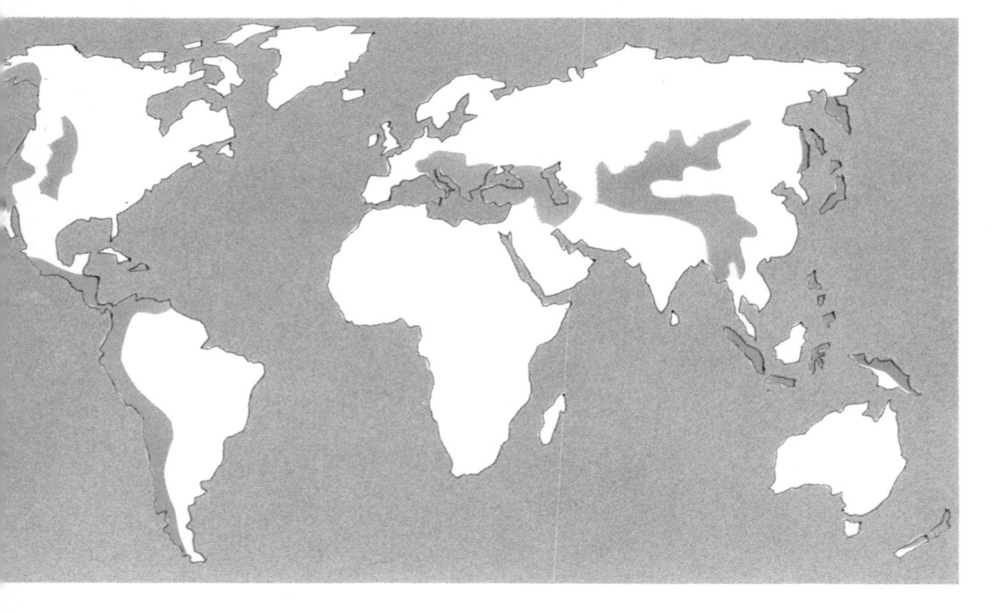

Perhaps as many as half a million earthquakes occur every year, but most are so slight that they are detected only by sensitive instruments, called *seismographs*. The most severe earthquakes occur in zones that follow the edges of the plates. The huge Armenian earthquake in 1988 resulted from the movement of plates.

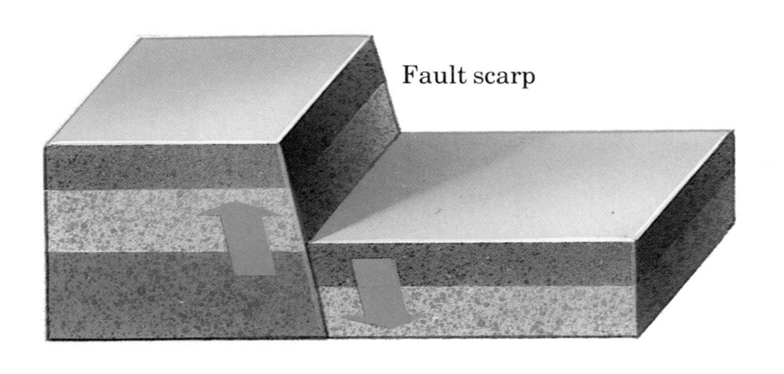

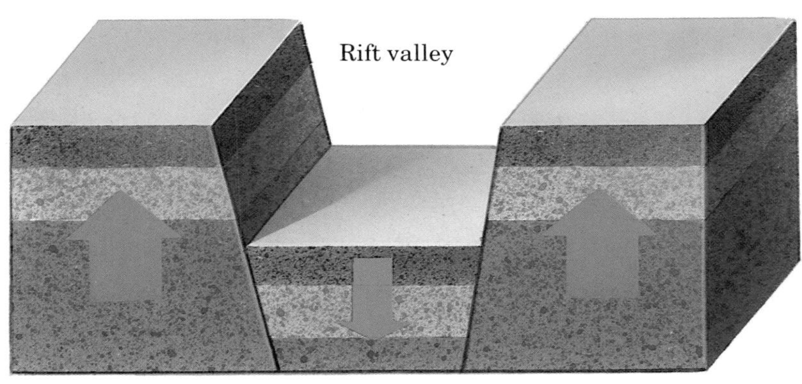

Sometimes rocks move past each other (above) along *transform* faults. Movements along the San Andreas fault near the coast of California caused a vast earthquake in 1906, and small tremors occur regularly. Buildings in the area are built to withstand another severe quake, which geologists predict may occur at any time.

as the San Andreas fault in California, two plates are sliding past each other. A sudden movement along the fault caused a severe earthquake in San Francisco in 1906. Many buildings collapsed. Others were destroyed by the fires that raged after the quake.

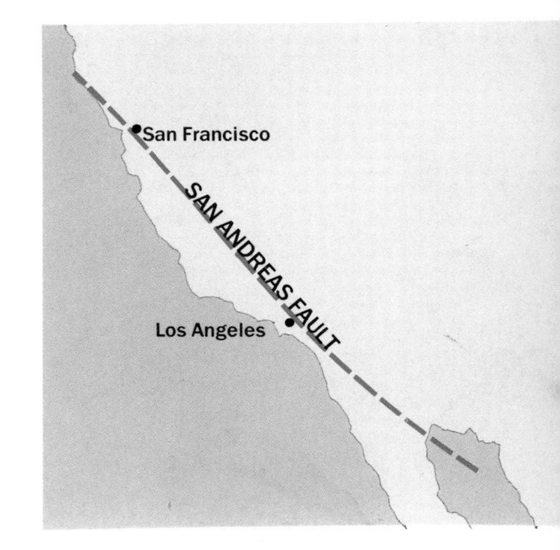

Rifts and Blocks

The tugging and pushing movements of plates crack rocks, creating huge faults. These movements cause blocks of land to sink or to rise up along fault lines. Sinking produces steep-sided valleys, such as the East African Rift Valley. Uplift produces block mountains, such as California's Sierra Nevada range.

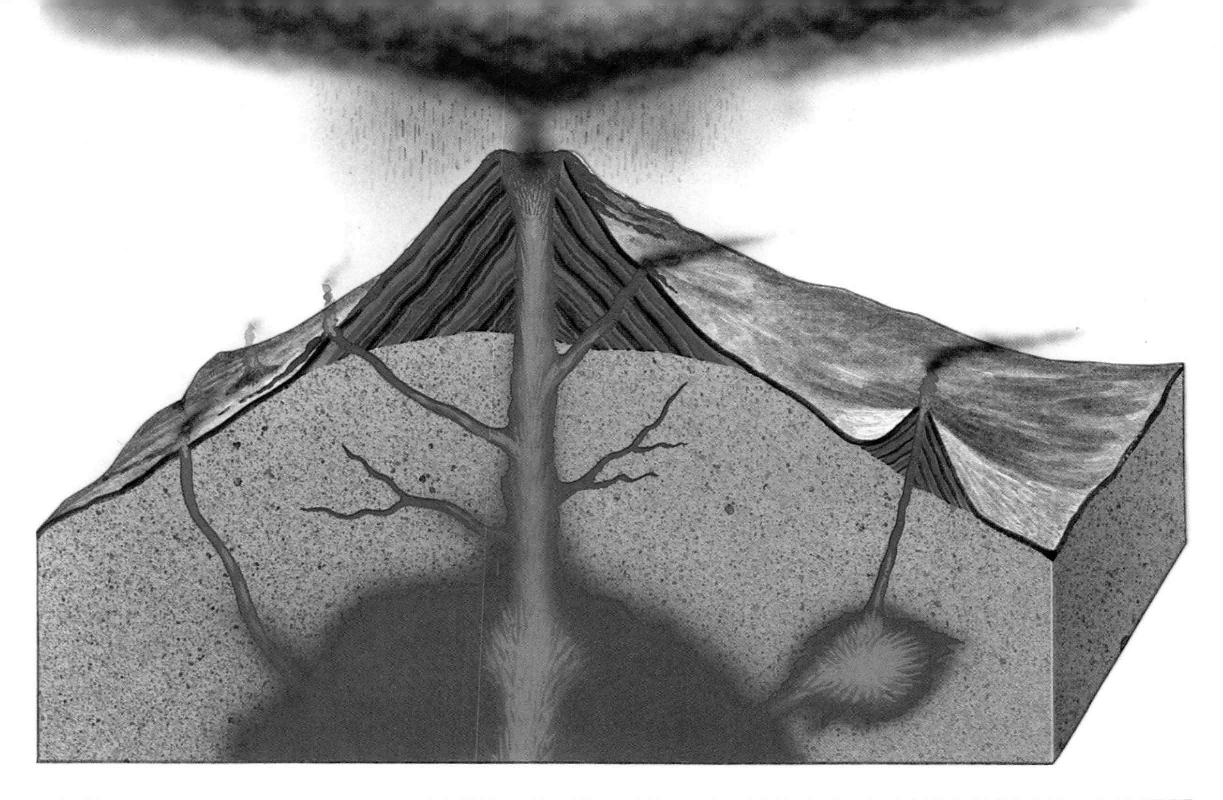

Active volcanoes may erupt explosively, hurling ash (shattered magma) into the air. Quiet eruptions are not explosive. Instead, streams of lava flow from the vent. Many volcanoes sometimes erupt explosively and sometimes quietly. As a result, the mountain that forms around the vent is made up of alternate layers of ash and hardened lava.

Making Mountains

In January, 1980, there was a vast explosion in the state of Washington in the northwest. Mount St. Helens volcano had erupted. A plug of solid lava was blown out of the top of the mountain, and a huge cloud of dust and ashes rose into the sky. What had happened?

Mount St. Helens lies near the edge of a plate that is being pushed under the continent. The magma which built up in the volcano was formed by the melting plate. In time, the pressure of the hot gas from the molten rock blew the top off the mountain. Other volcanoes are not so violent. When they erupt, hot lava

Mount St. Helens blows its top!

pours out of the vent, cools and solidifies. In time, a large cone-shaped mountain builds up.

A few volcanoes, including those in Hawaii, lie far from plate edges. Geologists believe that these volcanoes lie above a "hot spot" – a source of radioactive heat in the mantle.

Fold Mountains

Most of the mountains on land are formed by plate movements. When two plates carrying continents collide, the edges of the continents are buckled up, or folded, into huge mountain ranges. The Himalayas, the world's greatest mountain range, formed this way when a plate carrying India collided with one carrying Asia. Today, these land masses are fused together. The Ural Mountains also represent the border between what were, long ago, two separate land masses, Europe and Asia.

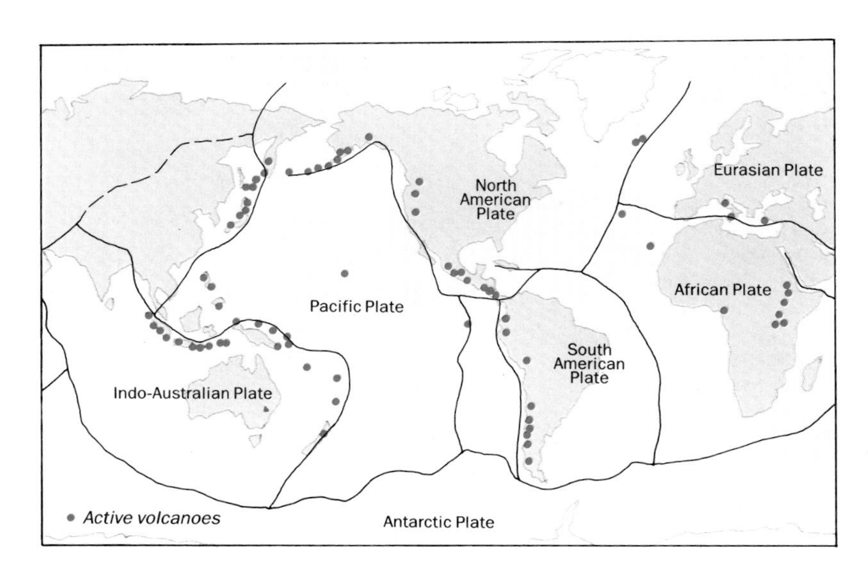

Most of the world's active volcanoes lie near plate edges, on the ocean ridges, or alongside deep ocean trenches. So many volcanoes encircle the Pacific Ocean that they are called the "Pacific ring of fire".

The Changing Earth

During its long history, the face of the world has continually changed. Moving at speeds of a few inches a year, the plates have shifted the oceans and continents. Some plates have collided and joined together; others have moved apart. As they have done so, new rock has formed, and the ocean above has widened, or a new ocean has been formed.

About 420 million years ago, there were four land masses. One, called Gondwanaland, consisted of all of the present-day southern continents. Two of the others (North America and Europe) collided, throwing up the Appalachians in North America and mountains in Greenland, Scandinavia and the British Isles. The new continent then collided with Asia, and the Urals were formed. North America, Europe and Asia now formed a single land mass, called Laurasia.

The maps show how the continents have moved during the last 200 million years. Before that time, there were several separate continents, which then joined together as Pangaea. This mass split again, and gradually the continents moved to their present positions. If the plates continue to move, the Atlantic Ocean will be far wider in 50 million years time, and Australia will be farther north, pushing up against Asia.

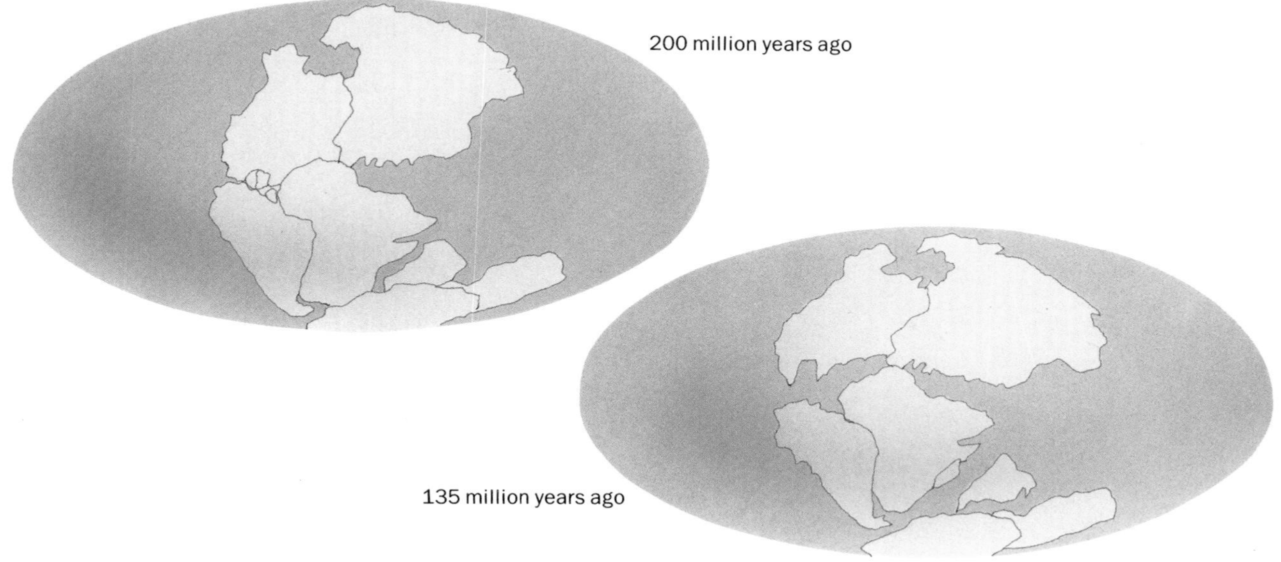

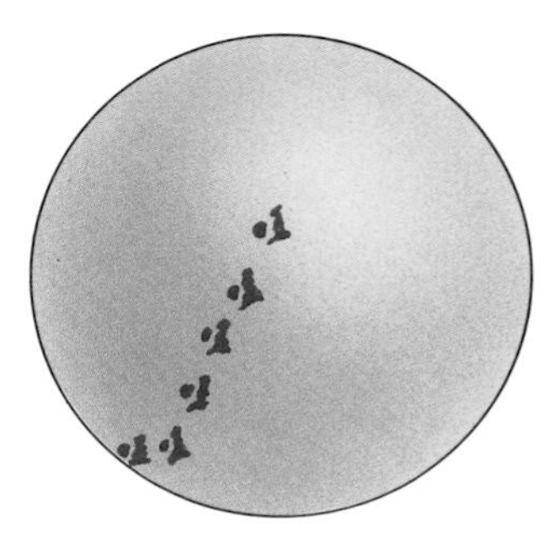
The British Isles are part of the continent of Europe. They have drifted – with the continent – 5,600 miles (9000 km) in the last 500 million years. The presence of coal means that at one time Europe was probably in the tropics. The ancient forests from which coal formed only grew in hot regions.

Pangaea to the Present Day

By 200 million years ago, Laurasia had joined onto Gondwanaland. Now there was just one land mass, called Pangaea. Around 140 million years ago, Pangaea started to break up, and the South Atlantic Ocean began to form between South America and Africa. Some 40 million years later, a gap between Europe and North America appeared and gradually grew into the North Atlantic Ocean. India, Australasia and Antarctica began to drift away from Africa.

About 50 million years ago, the plate bearing India began to push against the Eurasian plate, and the Himalayas began to form. They are still being uplifted to this day. Other plate movements are also going on. The Red Sea is getting wider as new crustal rock is being formed along its center. Eventually, this movement will close up the Persian Gulf, and a new fold mountain range will form.

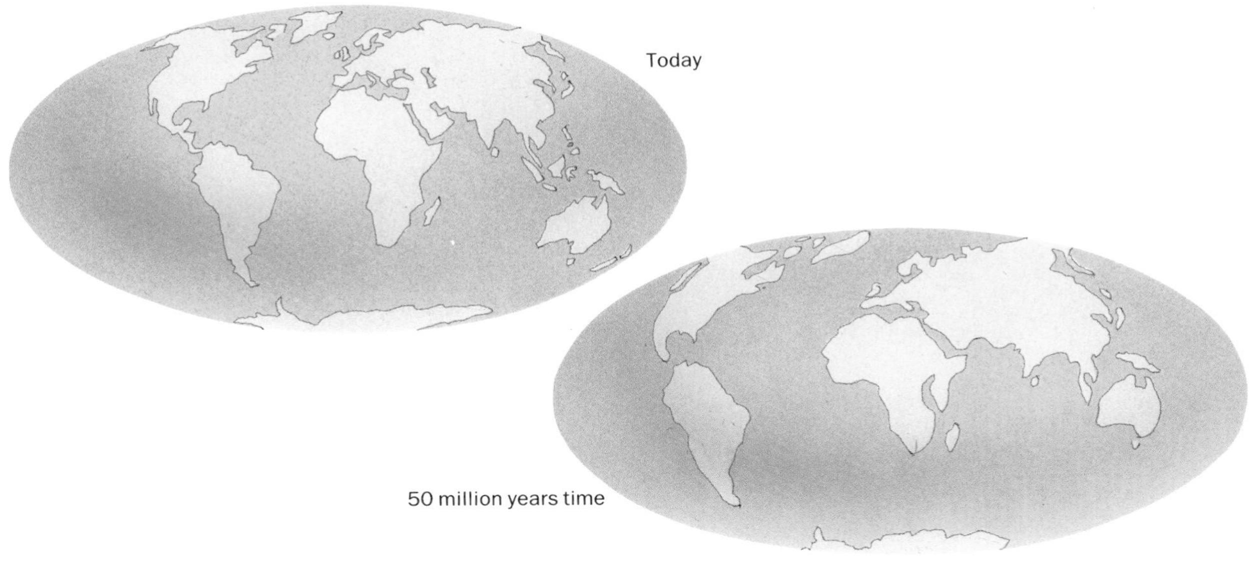

A Typical Continent

A typical continent is made up of several parts. The oldest parts are called *shields*. These generally low-lying areas contain the oldest rocks in the continent. Most of them are the remains of extremely old mountain ranges that have been worn almost flat. The shields contain igneous and sedimentary rocks. But these rocks have been so twisted, squeezed, cracked and re-heated that many of them have become metamorphic rocks.

Much of the shield is hidden by younger sedimentary rocks, many of them formed from fragments of material worn from ancient mountains. These areas, which are continuations of the shields, are called *stable platforms*. They have remained stable, or unchanged, for many millions of years. The vast flat

The diagram shows North America cut through to show what it is made of. In the west, new mountains are being formed as a plate pushes under the coast, causing earth movements and volcanoes. Inland, there are older ranges of mountains – the Rockies – created by the collision of two continents long ago. In the center is the shield – ancient rock which is now mostly covered by

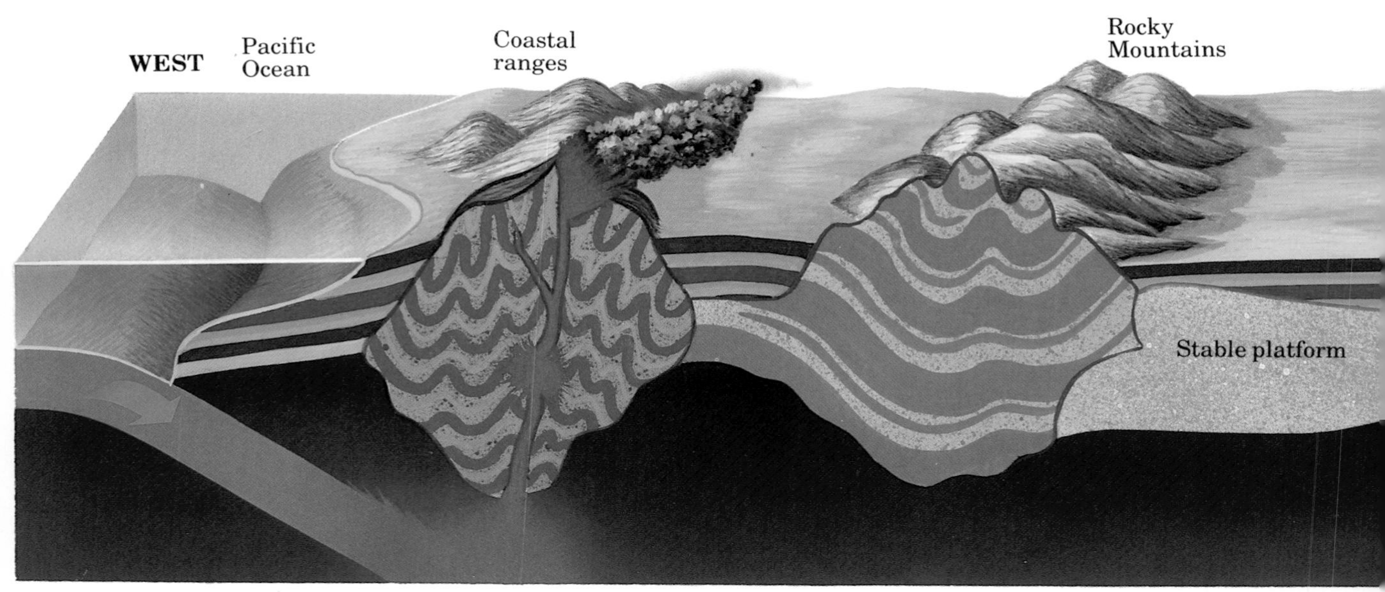

plains of the central U.S. are stable platforms.

Mountain ranges bordering the shields are more recent in origin. They are formed by the sideways pressure caused when two land masses are pushed against each other. The edges of the plates buckle up into fold mountain ranges, like the Himalayas. The Appalachians were once as high as the Himalayas, but they have been worn down since by rain, frost, ice and rivers. In the western U.S., on the edge of the continent, the young Cascade Mountains are still being formed.

Volcanoes occur on the sides of some continents. There, two plates are pushing against each other, but one plate is being forced under the continent. The edge of the descending plate melts, and some of this magma rises and emerges through the volcanoes.

sedimentary rocks, but is exposed in some places. In the east, a chain of old fold mountains reveals where two plates collided and the continents joined together. The mountains may once have been as high as the Himalayas, but they have been worn down by erosion. The eastern seaboard of the continent is the remains of the other continent which later broke away.

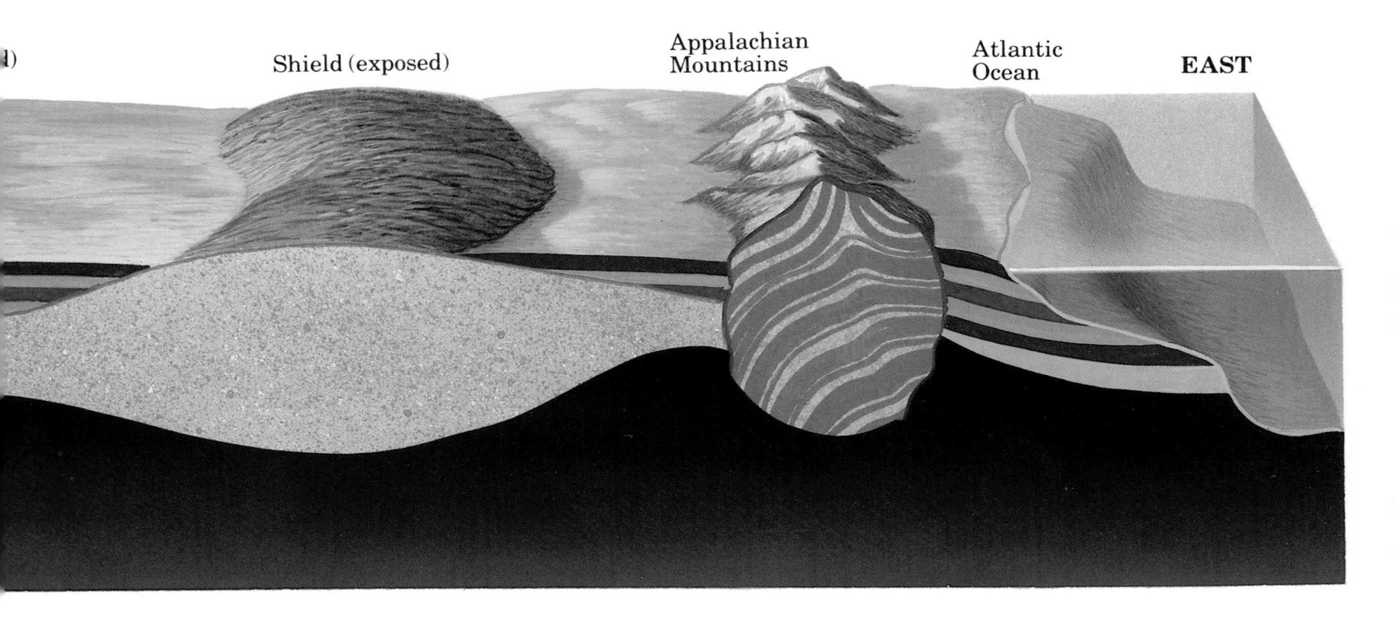

Life on the Continents

Living things evolve. They change in appearance and behavior in response to changes in their environment, such as climate, food supply, and location. The changes do not take place quickly, but slowly, like the movement of the continents.

Successful animals breed and spread. Fossils show how widespread many animals were when there was a single land mass. Where there are barriers, such as mountains or oceans, land animals cannot travel. So different kinds of animals evolve in different places. The mammals of Australia, which was cut off from the rest of the world a hundred million years ago, carry their young in pouches, and some kinds even lay eggs. Elsewhere, mammals give birth to live young.

Naturalists divide the world into *zoogeographical regions.* Each region contains animals that are found nowhere else. They have evolved since Pangaea broke up. Oceans, high mountains and deserts are barriers that mean that the land animals cannot travel from one region to another.

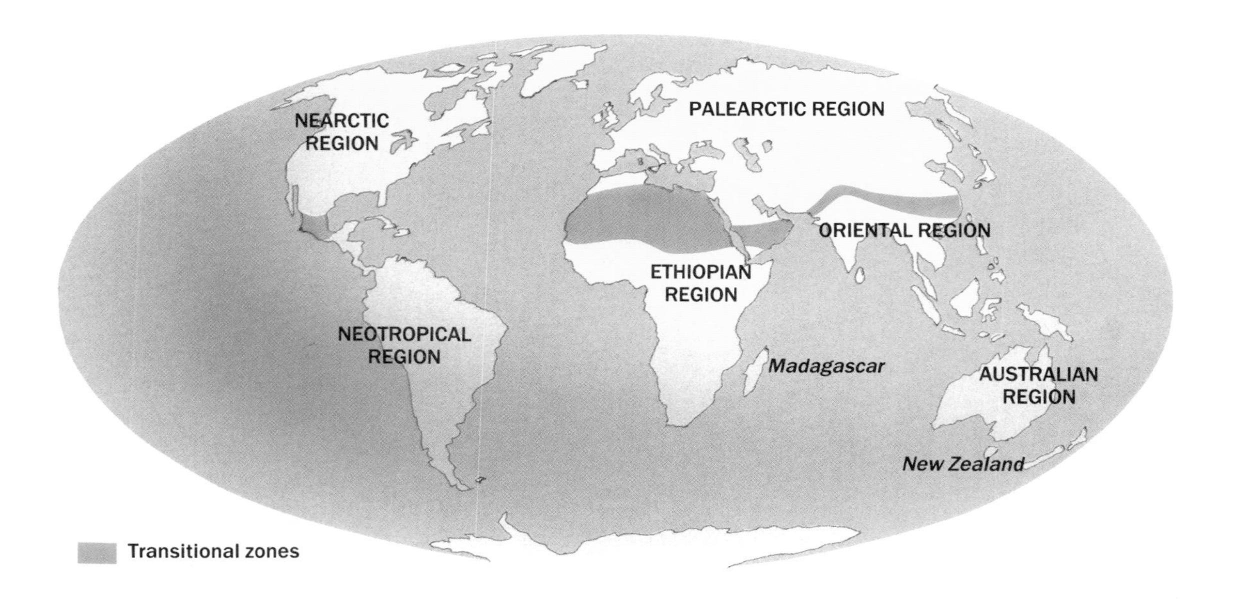

Animals that have evolved thousands of miles apart from each other may look the same because they live in the same sort of conditions. The emu and the ostrich are flightless birds whose feet are specially adapted to run. The kangaroo rat and the jerboa both live in deserts, while the deer and the antelope live on grasslands.

As the continents separated, the animals "on board each life raft" adapted to new environments – colder or warmer climates and different kinds of plants to eat.

Where conditions were similar, some animals on widely-separated continents grew to look and behave like each other. The grazing animals of Europe are deer. In Africa, they are antelopes. The antlers of one and the horns of the other are grown quite differently, but they look similar and have the same function. The little kangaroo rat of North American deserts has big ears that radiate heat. So do the jerboa and the gerbil of African and Asian deserts, and so does the pouched kangaroo mouse of Australia. These animals have all evolved in similar ways, even though they are not related.

Continent Profiles

NORTH AMERICA (including Central America)

Land The Canadian Shield covers much of eastern Canada north of the Great Lakes. To the west lie the interior plains, which extend south between the Rocky Mountains and the Appalachians. West of the Rockies are other ranges, high plateaus and basins. There are active volcanoes in southern Alaska, the northwestern U.S. and Mexico. Most Caribbean islands are volcanic in origin.

Climate North America includes the world's second largest ice sheet (Greenland), and much of northern Canada has an Arctic climate. Much of the rest of North America has cold winters and warm summers. The interior plains are dry grasslands; deserts occur in the southwestern U.S. and northern Mexico. Central America and the Caribbean have a mainly tropical, rainy climate.

Plants South of the icy Arctic lands is treeless tundra. To the south are forests of conifers, which merge into mixed forests, containing such deciduous trees as maple and beech. Redwoods, cedars, firs and spruce grow in the northwestern U.S., while hickory and oaks grow in the southeast. Tropical forests grow in parts of Central America.

Animals Caribou, polar bears and seals live in the north. Forest animals include black bears, deer and moose. The deserts contain rattlesnakes and many kinds of lizards. Alligators live in the southeast. The rain forests in the south contain many birds, monkeys and jaguars.

People The first people were Amerindians, who arrived more than 20,000 years ago. From the early 16th century, Spaniards settled Central America, where Spanish is now the main language. British and French people settled the north. French and English are official languages in Canada. English is the official language in the United States.

Economy North America is rich in natural resources; Canada and the U.S. are two of the world's richest nations. Mexico and the countries of Central America and the Caribbean are developing nations.

North America in Brief
Area 9,360,000 sq mi (24,249,000 sq km).
Population 418,000,000.
Independent countries 23.
Mountain ranges Alaska, Appalachians, Cascade, Rockies, Sierra Nevada, Sierra Madre.
Rivers Mississippi-Missouri, Mackenzie-Peace, Yukon, St. Lawrence.
Lakes Superior, Huron, Michigan, Great Bear.
Islands Greenland, Baffin, Victoria, Ellesmere and Newfoundland in the north. Cuba and other Caribbean islands in the south.

SOUTH AMERICA

Land The Andes Mountains run for more than 5,000 mi (8000 km) through western South America. Many peaks are volcanoes. Many rivers, such as the Amazon and Orinoco, rise in the Andes. Tributaries of the Amazon also rise in the Brazilian and Guiana highlands. The Guiana Highlands contain Angel Falls, the world's highest waterfall. Broad plains lie east of the Andes in the south.

Climate The equator runs through the hot, wet Amazon basin. Deserts run down the west coast from Ecuador to northern Chile (the Atacama Desert). The Brazilian Highlands are hot and dry. Much of southern South America has a moist temperate climate, but Patagonia is a cold desert.

Plants Tropical rain forest called *selva* covers much of the Amazon basin. Tropical grasslands lie to the north and south of the Amazon basin. The pampas of Argentina are temperate grasslands. The vegetation in the Andes varies according to the height, from tropical rain forest to peaks permanently covered with ice.

Animals The rain forests have a rich wildlife, including anacondas, anteaters, armadillos, capybaras, monkeys and many kinds of birds. The vicuña and guanaco are found in the high Andes, together with llamas and alpacas.

People The first people were Amerindians; in the 16th century, Europeans seized most of the continent. Spanish is the official language in most countries, with Portuguese in Brazil. English is spoken in Guyana, Dutch in Surinam, and French in French Guiana.

Economy South America is a developing continent, though Brazil and Argentina are expected to emerge as major powers in the next century.

South America in Brief
Area 6,884,300 sq mi (17,835,000 sq km)
Population 287,000,000.
Independent countries 12.
Mountain ranges Andes, Brazilian and Guiana Highlands.
Rivers Amazon, Rio de la Plata-Paraná, Madeira, Purus.
Lakes Titicaca in Peru and Bolivia.
Islands Tierra del Fuego, Falkland Islands (Britain), Galapagos Islands (Ecuador).

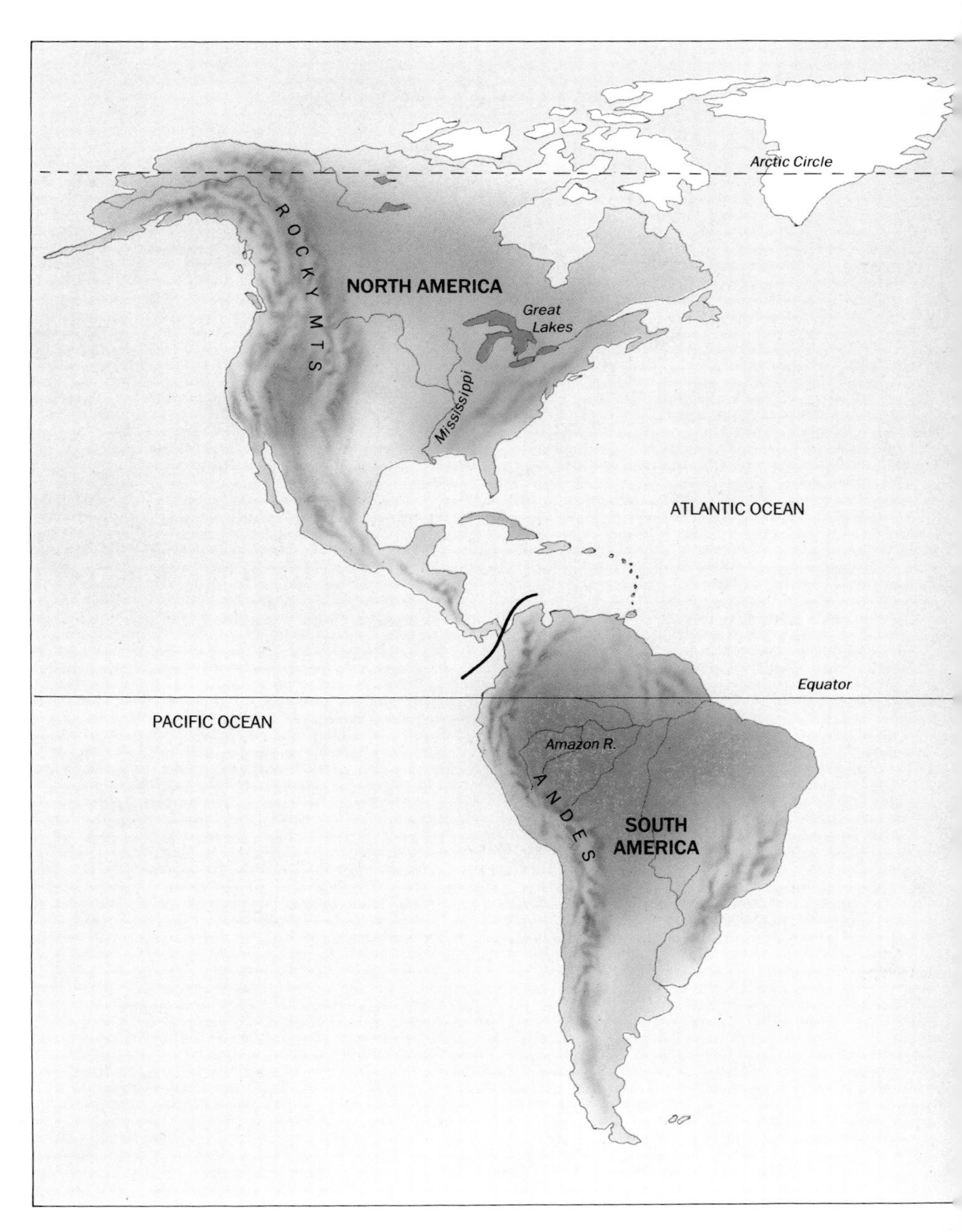

ASIA

Land Asia, the largest continent, contains many mountain ranges, including the world's highest, the Himalayas. To the east and southeast is an island zone, stretching through Japan, the Philippines and Indonesia, where the crust is unstable and volcanic eruptions and earthquakes are common. Asia also has some of the world's most densely populated river valleys and deltas. They contrast with the deserts of the southwest and center (the cold Gobi and Takla Makan deserts). Another desert (the Thar desert) is on the India/Pakistan border.

Climate The climate varies from bitterly cold Arctic conditions in the north to hot and steamy equatorial lands in the southeast. The southwest is hot and dry. South Asia is affected by monsoons – winds that blow from different directions according to the season. The summer monsoon in southern Asia often brings so much rain that floods occur.

Plants Asia has tundra and coniferous forests in the north and mixed and deciduous forests in the center. Dense tropical monsoon forest grows in the southeast, but plants are rare in the deserts.

Animals Foxes, lemmings and reindeer live in the far north, while brown bears, lynxes, otters and sable roam the northern forests. Southwest China is famed for the giant panda, but the greatest number of animal species live in the hot, wet lands of south, and southeastern Asia. They include crocodiles, rhinoceroses and tigers. Camels, oryxes and other animals live in the deserts.

People Asia was the home of early civilization and the birthplace of all the world's major religions. Today, it has many ethnic, language and religious groups. Each region has a distinct culture and its own characteristic political systems and traditions. China and India are the world's two most populous nations. They contain about 37 percent of the world's people.

Economy Except for Japan, one of the world's leading industrial powers, and the U.S.S.R., most of Asia consists of developing countries, though some are rich in resources, such as oil. But most Asians are poor compared with people in North America and Europe.

Asia in Brief
Area 16,963,500 sq mi (43,947,000 sq km), including 75 percent of the U.S.S.R., most of Turkey and the Sinai peninsula in Egypt.
Population 3,069,000,000.
Independent countries 42.
Mountains Elburz, Himalayas, Karakoram, Tien Shan, Zagros.
Rivers Yenisey, Yangtze, Ob-Irtysh, Huang He, Lena, Amur, Mekong, Ganges-Brahmaputra, Indus, Salween, Tigris-Euphrates.
Lakes Aral, Baikal, Balkhash.
Islands Borneo, Sumatra, Honshu, Sulawesi, Java.

EUROPE

Land East of the Scandinavian mountains in the north is the Baltic Shield, an area of ancient rocks and many lakes, which makes up most of Sweden and Finland. Iceland, which is also part of northern Europe, is a young volcanic island. The Great European Plain runs from southern Britain across central Europe and includes most of the European part of the U.S.S.R. The central uplands includes low mountains and plateaus, including the Meseta of Spain and Portugal, and the Massif Central in France. The Alpine mountain system includes not only the Alps, but also the Sierra Nevada, the Pyrenees, the Apennines in Italy, the Balkans and Carpathians in eastern Europe, and the Caucasus range in the U.S.S.R.

Climate Northern Europe lies in the Arctic, but coastal areas are warmed by an ocean current called the North Atlantic Drift. This current gives west-central Europe a mild, moist climate, but conditions become increasingly severe to the east. The warmest lands are in the south, and the Mediterranean Sea is a major resort area.

Plants The north contains treeless tundra and coniferous

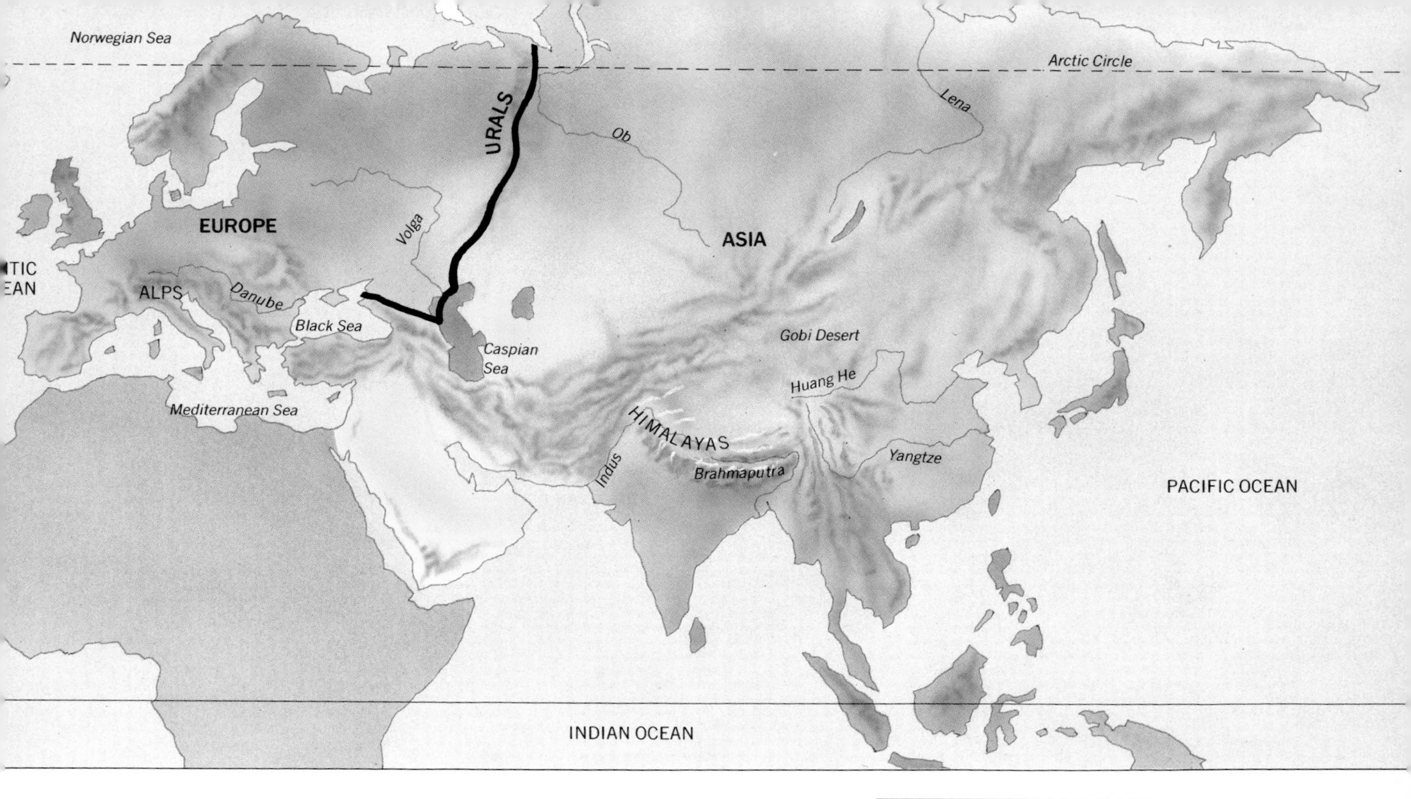

forest. The former deciduous forests of central Europe have been cut down. The most extensive grasslands were in the south-western U.S.S.R., but they are now largely used for crops. The Mediterranean region has much scrubland, called *maquis*, with evergreens such as cork oaks and olive trees.

Animals Animal life has been much reduced by human activities. Brown bears and reindeer live in the north. The Alpine mountains contain goat-like chamois and ibex. More common are such animals as badgers, moles, otters, rabbits, squirrels and many kinds of birds.

People Europe is a densely populated continent, the home of the Greek and Roman civilizations whose influence can still be seen. About 50 languages and many more dialects are spoken. Europe led the way in exploring and colonizing the world. Today, several European countries are home to sizeable groups of people of African and Asian origin.

Economy Europe's countries are highly developed, with many manufacturing industries and highly productive farms. As a result, most people enjoy high standards of living compared with those of people in developing countries.

Europe in Brief
Area 4,065,350 sq mi (10,532,000 sq km), including 25 percent of the U.S.S.R. and 3 percent of Turkey.
Population 692,000,000.
Independent countries 34.
Mountain ranges Alps, Apennines, Balkans, Carpathians, Caucasus, Pyrenees, Sierra Nevada.
Rivers Volga, Danube, Don, Rhine.
Lakes Caspian Sea, Ladoga.
Islands Great Britain, Iceland, Ireland.

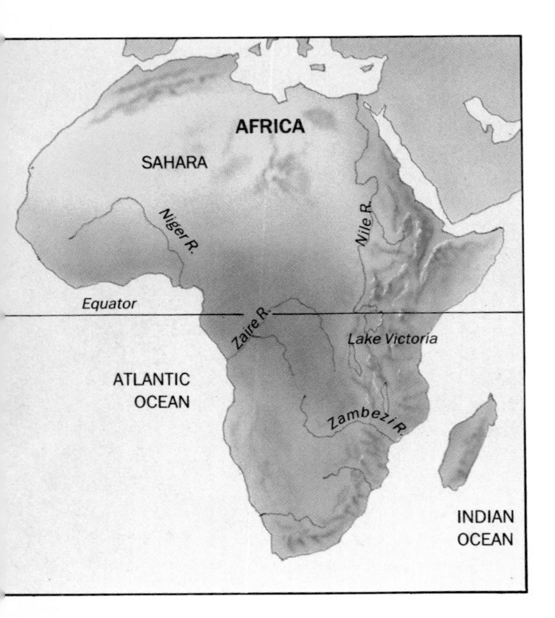

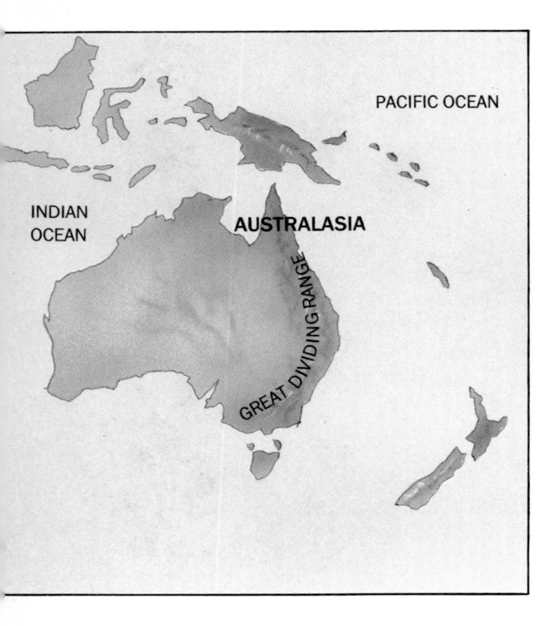

AFRICA

Land Most of Africa is a high plateau made of ancient shield rocks. The Drakensberg in South Africa is the uptilted rim of the plateau, but the highest peaks, Kilimanjaro and Kenya, are extinct volcanoes. East Africa contains the world's longest rift valley, which contains many lakes. Africa's long rivers are important for transport, but near the coast (at the edge of the plateau), waterfalls and rapids make them unnavigable.

Climate Africa is a warm continent bisected by the equator. Between the warm Mediterranean regions in the north and the far southwest (around Cape Town) are: hot deserts – the Sahara (the world's largest), the Somali, the Namib, and the Kalahari; tropical regions with wet and dry seasons; and equatorial regions, which are hot and wet throughout the year.

Plants The Mediterranean regions contain much scrubland, while few plants grow in the Sahara, except date palms around scattered oases. Tropical grassland (*savanna*) is widespread, and rain forests grow in the hot, wet equatorial regions. The Kalahari in southern Africa has coarse grasses and scrub, while the interior of South Africa is *veld* (dry grassland).

Animals Africa is rich in animal species, especially in the rain forests and savanna. Forest animals include chimpanzees and gorillas, and crocodiles and hippopotamuses bask in the rivers. Savanna animals include buffaloes, elephants, giraffes, leopards, ostriches and zebras.

People The two main groups are the Arabs and Berbers in the north, and the black Africans of central and southern Africa. Until about 1960, most of Africa was ruled by European countries, but nearly all of Africa is now independent. The largest group of European descendents is in South Africa.

Economy Most African countries are poor and underdeveloped compared with those of Europe and North America, though Africa has many mineral resources. The rainfall is unreliable in many areas. Droughts cause crop failures and famines.

Africa in Brief
Area 11,707,380 sq mi (30,330,000 sq km)
Population 600,000,000.
Independent countries 51.
Mountains Ahaggar and Tibesti massifs in the Sahara; the folded Atlas Mountains; the Ruwenzori range in East Africa; and the Drakensberg in South Africa.
Rivers Nile, Zaire, Niger, Zambezi, Orange, Limpopo.
Lakes Victoria, Tanganyika, Malawi, Chad.
Islands Madagascar.

AUSTRALASIA

Land Australia is an ancient land mass which has been so worn down that it is the flattest continent. The chief highlands are the Great Dividing Range in the east. Australia contains the world's largest coral formation, the Great Barrier Reef, off the northeast coast. New Guinea is mountainous, with the region's highest peaks. New Zealand's Southern Alps are on South Island. North Island contains active volcanoes.

Climate Australia is the driest continent, and much of west and central Australia is desert or semi-desert. The sunny east and southeast coasts have ample rain and contain most of the country's population. Northern Australia and New Guinea have a tropical, rainy climate. New Zealand has a mild, moist climate.

Plants New Guinea has rain forests and tropical grassland. Savanna covers parts of northern Australia, while the west is largely desert or dry grassland. Acacia and eucalyptus are common shrubs and trees. New Zealand has forests of evergreens and ferns.

Animals: New Guinea has many animals, including crocodiles, snakes and brightly colored birds. Australia is famed for its marsupials, including kangaroos, koalas and wallabies. Platypuses and echidnas are mammals that lay eggs. Many of New Zealand's animals have been introduced. Native species include reptiles called tuataras and many birds, including kiwis and keas.

People Most people in New Guinea are Melanesians. The aboriginal people of Australia settled there at least 40,000 years ago. But most Australians are descendants of immigrants from Europe, especially the British Isles. The Maoris of New Zealand are Polynesians, but most New Zealanders are of British origin. English is the official language in three countries of Australasia.

Economy Papua New Guinea is a developing country, but Australia, with its huge mineral resources, and New Zealand, famed for its agricultural exports, are both prosperous, and the people have a high standard of living.

Australasia in Brief
Area 2,965,250 sq mi (7,682,000 sq km) (Aus); 311,656 sq mi (807,400 sq km) (NG); 103,856 sq mi (269,057 sq km) (NZ).
Population 16,297,000 (Aus); 4,720,000 (NG); 3,372,000 (NZ).
Independent countries 3.
Mountains Great Dividing Range (Aus); Owen-Stanley (NG); Southern Alps (NZ).
Rivers Murray-Darling (Aus).
Islands New Guinea, North Island and South Island (NZ), Tasmania.

ANTARCTICA

Ice and snow cover about 98 percent of Antarctica, the world's coldest continent. The ice reaches a depth of 15,800 ft (4800 m), but in places, mountain peaks appear through the ice. Mount Erebus, an active volcano, is on Ross Island. Wildlife is confined to the edges of the continent and includes penguins and seals. Whales swim in the surrounding seas. The only people are scientists who spend periods working there.

Antarctica in Brief
Area 5,404,000 sq mi (14,000,000 sq km)
Mountains Transantarctic Ellsworth, Antarctic Peninsula.

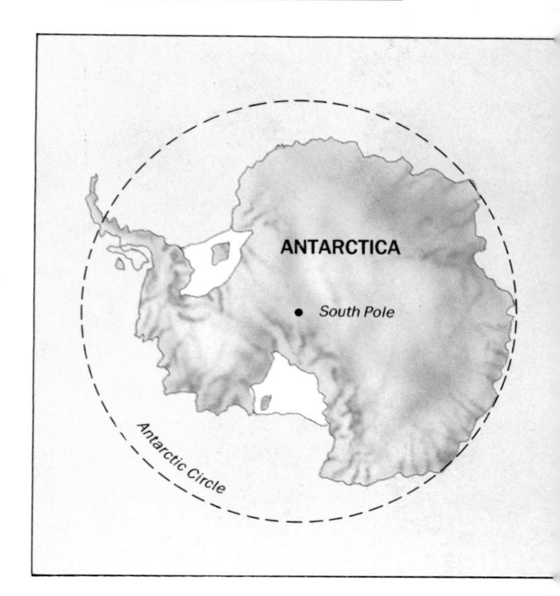

Index

Descriptions and statistics of individual continents are listed under Continent Profiles on pages 29-31.